Statistics is Easy

Case Studies on Real Scientific Datasets

Synthesis Lectures on Mathematics and Statistics

Editor
Steven G. Krantz, *Washington University, St. Louis*

Statistics is Easy: Case Studies on Real Scientific Datasets

Manpreet Singh Katari, Sudarshini Tyagi, and Dennis Shasha

ISBN: 978-3-031-01305-8   paperback
ISBN: 978-3-031-02433-7   ebook
ISBN: 978-3-031-00279-3   hardcover

DOI 10.1007/978-3-031-02433-7

A Publication in the Springer series
SYNTHESIS LECTURES ON MATHEMATICS AND STATISTICS

Lecture #39
Series Editor: Steven G. Krantz, *Washington University, St. Louis*
Series ISSN
Print 1938-1743   Electronic 1938-1751

Statistics is Easy

Case Studies on Real Scientific Datasets

Manpreet Singh Katari
New York University

Sudarshini Tyagi
Goldman Sachs

Dennis Shasha
New York University

SYNTHESIS LECTURES ON MATHEMATICS AND STATISTICS #39

ABSTRACT

Computational analysis of natural science experiments often confronts noisy data due to natural variability in environment or measurement. Drawing conclusions in the face of such noise entails a statistical analysis.

Parametric statistical methods assume that the data is a sample from a population that can be characterized by a specific distribution (e.g., a normal distribution). When the assumption is true, parametric approaches can lead to high confidence predictions. However, in many cases particular distribution assumptions do not hold. In that case, assuming a distribution may yield false conclusions.

he companion book *Statistics is Easy!* gave a (nearly) equation-free introduction to non-parametric (i.e., no distribution assumption) statistical methods. The present book applies data preparation, machine learning, and nonparametric statistics to three quite different life science datasets. We provide the code as applied to each dataset in both R and Python 3. We also include exercises for self-study or classroom use.

KEYWORDS

scientific data, case studies, nonparametric statistics, machine learning, data cleaning, null value imputation

Readers can find Statistics is Easy! Second Edition by Dennis Shasha and Manda Wilson at https://doi.org/10.2200/S00295ED1V01Y201009MAS008 or http://bit.ly/EasyStats2.

Contents

Acknowledgments

First, we would like to thank our students whose questions over the years have shaped the narrative of this book. Second, we'd like to thank Professor Kristin Gunsalus for her careful review and thoughtful comments. Third, we'd like to thank the editorial and publishing team at Morgan & Claypool: Diane Cerra, Christine Kiilerich, Sara Kreisman, C. L. Tondo, Laurel Muller, Brent Beckley, and Sue Beckley. Last and certainly not least, we'd like to thank our families for enduring our absences while we worked on this book, which, though concise, required sustained effort over two years.

Shasha has been partially supported by NIH 1R01GM121753-01A1 the U.S. National Science Foundation under grants MCB-1412232 and IOS-1339362 and MCB-0929339. Katari has been partially supported by NIH-NIGMS: 1 R01 GM121753-01 and DOE-BER: DE-SC0014377. This support is greatly appreciated.

A special acknowledgment to Rohini, Anjini, and Reena Katari for their continued support and inspiration.

Manpreet Singh Katari, Sudarshini Tyagi, and Dennis Shasha
February 2021

CHAPTER 1

Introduction

The book is aimed at computer scientists, bioinformaticians, and computationally oriented scientists. We describe three case studies in order to illustrate different aspects of data analysis. All code is provided in both R and Python 3 and is available on Github (https://github.com/StatisticsIsEasy/StatisticsIsEasy and https://github.com/StatisticsIsEasy/CaseStudies).

In the Basic Workflow section of the present chapter, we describe the steps of any data-driven study: data preparation, analysis (including machine learning), and statistical evaluation.

In the Technology Choices section, we describe alternative methods to achieve the objective of each step of the analysis workflow.

Finally, in the Case Studies Overview section, we briefly describe each case study, its dataset, and the techniques applied for each of its workflow steps.

Faced with a dataset and an analysis goal of your own, we suggest that you find the most similar case study from this book and use our code as a starting template to solve your problem. You may then add and substitute code from elsewhere onto that basic scaffold.

1.1   BASIC WORKFLOW

The basic data-driven workflow consists of three parts.

1. **Data preparation**—Put the data in good form for analysis. This constitutes the majority of the work for a practicing data scientist and consists of the following general steps.

   (a) Format the data used in the dataset, usually through some form of parsing. The end result is often a table. For example, for the breast cancer dataset, each row corresponds to a patient and each column value is some tumor measurement.

   (b) Normalize raw data so that different features or experimental measurements are rendered comparable.

   (c) Handle missing data by ignoring some inputs or imputing missing values.

2. **Analysis**—Analysis can mean calculating some metric of interest or creating a causal and/or predictive model. The metric might be something simple like weight or yield. Causality and prediction models are built from inference methods such as regression, machine learning, or forecasting. They might be used to diagnose patients as having malignant or benign growths.

3. **Statistical Analysis**—The first step is to assess the analytical measurements and/or models to determine whether they could have arisen due to random chance. In the case of a measurement (say the weight of chickens), the question might be which treatment (e.g., chicken diets) might cause a statistically significantly different value of some measurement (e.g., final chick weight) than a standard diet. In the case of inference models, the question might be which model gives statistically significantly better classification results (e.g., in the breast cancer case, which model gives a more accurate diagnosis). Here are some typical choices and tasks in statistical analysis.

   (a) **The choice between a paired vs. unpaired statistical analysis**. For example, suppose we are evaluating a treatment, and that one group of patients receive the treatment and a different group does not (the control group). We must use an unpaired analysis, because they are different people. On the other hand, if the same patients are measured before and after treatment (as when sick patients are given a medicine in the cystic fibrosis study), then we can use a paired test, because we are comparing the same person before and after a treatment. A paired test is better at detecting subtle effects, because it removes the effect of confounding factors due to differences among patient populations.

   (b) **The choice of accuracy measure.** For classification problems, these are based on notions of "precision" (roughly, the fraction of predictions of the class of interest that are correct) and "recall" (roughly, the fraction of entities in the class of interest that are correctly predicted). For problems in which we are trying to predict a numerical value, the accuracy measure is based on either a relative or absolute numerical error of a predicted value to an actual value. We discuss both of these cases later in this chapter.

   (c) **The task of determining whether some causal factor (e.g., an experimental perturbation) might be associated with some change** in a measurement of interest more than would be expected from random chance alone. For example, the putative causal factor might be a change in diet and the measurement might be the final chicken weight. If the measurement with the factor (e.g., new diet) is different enough from the measurement without the factor (e.g., normal diet), then the difference would be said to have a low *p-value*. When the p-value is low enough, then random chance is unlikely to be an explanation of the difference.

   (d) **Choosing the correct multiple hypothesis testing correction procedure.** When one asks about many phenomena, e.g., which of thousands of genes is affected by cystic fibrosis, computing a p-value is not enough. The reason is that by random chance, some changes to genes that have a low p-value would materialize, even if health status had no real effect on genes.

The choice then is which multiple hypothesis testing correction procedure to use. The false discovery rate is the most common. The false discovery rate is an estimate of the fraction of genes that might fall below a certain p-value threshold simply due to chance (i.e., unrelated to health status).

(e) **Estimating the range of the effect** we can expect by choosing one course of action rather than another. In the chicken case study, for example, we might ask about the range of the expected magnitude of the weight gain for a chicken on a new diet compared with a chicken on a normal diet.

(f) **The determination of which features of some treatment or condition that most influence the result.** For example, in the breast cancer case study, we want to find the features of cell nuclei that are most diagnostic for breast cancer.

1.2 TECHNOLOGY CHOICES

As mentioned above, the overall workflow consists of data preparation, data analysis, and statistical evaluation. This section gives a brief review of the technologies for each of these. The github (https://github.com/StatisticsIsEasy/CaseStudies) associated with this book has all the code in this section and that we have used throughout the book, in R or Python 3 (and usually both). Alternatively, we authors can send you a zip file.

1.2.1 DATA PREPARATION TECHNOLOGIES

Before preparing the data, it's good to look at it. We mean that literally. Because the culture of computing is to care about method more than about data, computationally trained people don't stare at the data enough. For example, due to experimental methods, some data values may have wildly differing magnitudes, outliers, and/or null values. Eyeballing the data will often suggest actions to perform. For example, there might be outliers, which are values that are substantially different from mean values, indicating an error in data recording. Sometimes, a mechanical flaw in an instrument can render an entire experiment's results invalid.

Normalization

Most analyses will have to compare different features and weight their effect on some metric of interest. To take an example from normal life, a person's temperature is more important than his or her shoe size in determining whether that person feels sick. If the different features can take vastly different ranges of values, that may skew the weighting in any analysis.

Certain computational methods therefore perform *z scaling* of the values: For each value v of feature X, scaling subtracts the mean of X from v and then divides that result by the standard deviation of X. Scaling makes sure each value of X is transformed to a number of standard deviations away from the mean of X, resulting in what is known as the *z-score*.

$$z = \frac{v - \mu}{\sigma}.$$ (1.1)

For example, let's take 15 randomly generated body mass index (BMI) values[1]:

```
import numpy as np
bmi = np.random.randint(18,32,15)
bmi

array([30, 22, 21, 25, 25, 29, 23, 20, 25, 27, 20, 18, 29, 27, 19])
```

Now we calculate the mean and standard deviation of these array values. Then, for each value, we will subtract the mean and divide by the standard deviation to get the z-score scaled values.

```
mu = np.mean(bmi)
sigma = np.std(bmi)
(bmi - mu)/sigma

array([ 1.58851015, -0.52950338, -0.79425507,  0.26475169,  0.26475169,
        1.32375846, -0.26475169, -1.05900676,  0.26475169,  0.79425507,
       -1.05900676, -1.58851015,  1.32375846,  0.79425507, -1.32375846])
```

Different normalization methods can work better for different data. For example, when preparing each RNA-seq dataset for the cystic fibrosis analysis, we may be more interested in the relative amount of each gene's RNA compared to other genes rather than the absolute amount. Thus, normalization would consist of dividing each gene's counts by the total counts for a given RNA-seq reading. This calculates the proportion of the counts that belong to the gene for that reading. We do this for all RNA-Seq readings so they can now be comparable [Bushel, 2020].

Normalization, whether to do it and how to do it, can influence all downstream analysis, so it's good to see whether the results of an analysis are robust to different normalization techniques.

Missing Data

There are two basic ways to deal with missing data within datasets: (i) remove any data item containing missing data, or (ii) infer the missing values from other data (called *imputation*).

---

[1]Our github site has both R and Python code, but we use Python in the book text.

While removing the data item requires less effort and thought, it can sometimes lead to a loss of hard-to-obtain information. For example, suppose some patient in an Alzheimer's study is missing a blood plasma test during a particular visit. The data for that day would include other informative metrics measured for the patient. Throwing out the entire day's data because of the one missing test value might result in an excessive loss of information. Common imputation methods include the following.

1. **Method 1:** Replace the missing values of some measurement with the arithmetic mean value of that measurement. In the example above, if the blood pressure reading is missing for a patient for a single visit, imputation could simply take the arithmetic mean of the blood pressure readings of all other visits of that patient. If no other measurements are available for that patient, than take the median measurement of all patient measurements.

2. **Method 2:** Replace the missing values with median. As in the previous method, but use median instead of arithmetic mean.

3. **Method 3:** Linear interpolation. If data comes in the form of a time series per individual, then the value at time t of some measurement for individual may be well estimated by taking the arithmetic mean of the value of that measurements of that individual at times $t-1$ and $t+1$.

4. **Method 4:** Design a machine learning model to predict the value of a missing measurement given the values of other measurements. This entails building, for each measurement type with missing values, a model based on the values of other measurement types.

Which imputation method to use can have an important effect on the results of an analysis, so the method and its justification should be carefully explained in any research paper.

1.2.2 METHOD SELECTION

Different researchers may analyze the same dataset in different ways. For instance, in the genomic data example of cystic fibrosis, we might be looking for genes that are differentially expressed due to disease or that change due to some treatment. Alternatively, we might be interested in predicting whether a patient has cystic fibrosis or not. The objective of the study determines the inference methods to select.

Many books (e.g., Hastie [2001]) and thousands of papers have been written about inference methods. This book includes just a few that work well for modest amounts of data: regression methods, decision trees, random forests, and support vector machines. Neural networks generally are most useful when datasets are much larger than those considered here. As new methods become available, packages will incorporate them and you will be able to use them.

Quality Metrics for Classification

In addition to analytical methods for classification, there are metrics to measure the accuracy of the methods on the given data. These are based on two notions called Precision and Recall. For the sake of illustration, we describe these notions in terms of classification of cancer.

- Precision is the number of correct cancer classifications divided by all predicted cancer classifications. Symbolically, suppose CancAll is the set of people who have cancer; ClassCorrect is the set of people whom the classifier claims have cancer and do; and ClassAll is the set of people whom the classifier claims have cancer whether they do or not. Then:

$$precision = \frac{||ClassCorrect||}{||ClassAll||}.$$
(1.2)

  (Note that the $||S||$ notation means the number of members of set S.)

  Higher precision means few false positives.

- Recall is the number of correct cancer classifications divided by all patients who have cancer. Using the symbols from the previous bullet:

$$recall = \frac{||ClassCorrect||}{||CancAll||}.$$
(1.3)

  High recall means few false negatives.

- There is usually a trade-off between precision and recall. A low acceptance threshold may yield higher recall but lower precision and conversely. To balance the two, one can capture both in one metric called the F-score (also known as the F1-score).

  The F-score can be calculated by dividing the product of precision and recall by the sum of precision and recall and multiplying the result by 2.

  The formula for the F score is:

$$F\ score = 2\frac{(precision * recall)}{(precision + recall)}.$$
(1.4)

Quality Metrics for Prediction

When the goal is to predict a real number value (e.g., to predict the weight of a chicken), precision and recall would make sense only if we artificially discretize the weights into bins. This can sometimes be useful, but usually is not. Instead we look at a measure called RMSE (Root Mean Square Error), which is simply the square root of the arithmetic mean of the squares of the differences between each of the n predictions ($\hat{y}_i$ below) and the corresponding correct value (y_i):

$$\text{RMSE} = \sqrt{\frac{1}{n} \sum_{i=1}^{n} (\hat{y}_i - y_i)^2}.$$

1.2.3   STATISTICAL ANALYSIS TECHNOLOGIES

Once we have obtained the results from the experiment, we want to determine whether they indicate an effect beyond random chance and the size of that effect.

We use a nonparametric approach as in the companion book *Statistics is Easy*, because making a specific distribution assumption (e.g., that the data is normally distributed) may not be justified in many cases and may therefore lead to erroneous conclusions.

The only assumption of the nonparametric approach is that the data collected is an unbiased sample from the population. Usually, this means that the sample is taken randomly with uniform likelihood without replacement from an underlying population. This is a far weaker assumption than assuming that the whole population can be characterized by some particular distribution.

Establishing Significance: p-values

In many scientific settings, one is trying to determine whether some factor is critical to a result. In the chicken dataset described in the next chapter, for example, we want to determine whether diet influences the final body weight of the chicken. Because of the wide variability among chickens, any two groups of chickens will have different mean final weights even if the two groups follow the same diet. To evaluate whether some change in diet is likely to overcome this random fluctuation, we use the notion of p-value, briefly introduced above.

Suppose, for example, that an experiment reports that diet A gives a higher average weight than diet B. The *p-value* is the probability that the observed differences among the diet groups could have happened by chance. If the p-value is low (conventionally under 0.05), then the null hypothesis that the diet didn't matter is unlikely, so we may be justified in concluding that the diet likely did matter.

In parametric statistics, one can find a p-value by assuming a distribution and evaluating the observed results based on that distribution. Because our philosophy is to assume nothing about a distribution, we use a relabeling and counting technique called "shuffling" which is explained in the companion book *Statistics is Easy* (as are all the statistical techniques used here), though we'll review it later in the Chick Weight case study and also apply it in the Cystic Fibrosis case study.

Nonparametric Power Analysis: Estimating How Big a Sample Size Should Be

A p-value analysis that leads to feedback for the experimentalist in determining a sufficiently large sample size. Here is how. Suppose the p-value is very high. Resampling can find the number of replicates that might reduce the metric to a more desirable number.

An example will help explain this: suppose that a treatment is given to some set of patients T and a placebo to some set P. Suppose that the patients in T do in fact receive a greater benefit than the patients in P, but the result has an excessively high p-value. One can computationally create a new treatment set T' consisting of T and say some fraction f (this fraction could be a multiple, e.g., 3/2) of $||T||$ additional values that are drawn randomly and uniformly without replacement from T. One could use the same fraction f to create a new set P' from P. If as a result the difference in benefit between T' and P' falls below some p-value threshold, then a new experiment in which the treatment pool is of size $||T'||$ or more and the placebo pool is of size $||P'||$ or more may lead to a desired p-value if the effect is in fact real.

Magnitude of an Effect: Confidence Intervals

Suppose that we find that diet has some effect on chicken weight (i.e., there is a low p-value that diet has no effect). A natural next question is to determine how big that effect is.

Suppose that in the sample the average weight of chickens eating diet A is 0.8 kg more than the average weight of chickens eating diet B. We might want to know how different that average weight might be for a different group of chickens who take diet A compared to a group who take diet B. The result might say something like: we expect the average difference to lie between 0.68–0.85 kg 90% of the time. In that case, we'd say the 90% confidence interval lies between 0.68 and 0.85 kg.

We will use a nonparametric statistics technique called *bootstrapping* to compute confidence intervals. This is also discussed in *Statistics is Easy* and briefly reviewed in the Chick Weight study and also used in the Cystic Fibrosis case study.

Multiple Hypothesis Testing

For a single scientific question (e.g., does diet A lead to a higher weight than diet B), the p-value is an appropriate indicator of statistical significance. When there are multiple scientific questions (e.g., which genes have been affected by a treatment), the p-value is only the first step.

To use an everyday example, if we compare different diets for chickens along many measurement dimensions: final weight, number of clucks per minute, number of hours sleep, ... then for some measurement dimension the diet difference may show a low p-value just by chance. This is called the multiple testing effect: even very low p-values will occasionally appear for a measurement type even when a treatment has no real effect.

There are several ways to overcome the multiple hypothesis testing effect. The two we will consider in this book are (1) *Bonferroni correction* and (2) *false discovery rate (FDR)*, the latter of which we introduced briefly above.

The *Bonferroni correction* evaluates the probability that there is at least one false positive (sometimes called the family-wise error probability). Operationally, one takes some probability cutoff, denoted α, and divides it by the number of tests. For example, if $\alpha = 0.05$, and there are 10,000 tests, then the Bonferroni correction [Bonferroni, 1936] will accept only those test results that have a p-value of $\frac{0.05}{10,000}$ or less. The net effect is that the probability that one or more of those accepted tests had the observed value due to chance is 0.05 or less.

False discovery rate is defined as the number of measurement differences with a certain level of p-value which could have that level merely by chance (in our example, without any influence from the diet). Here (as in most current RNA-seq analyses) we use the technique of Benjamini and Hochberg [1995] though there are others. We describe that technique in the Cystic Fibrosis chapter.

1.3   OVERVIEW OF THE CASE STUDIES

For each case study, we describe the dataset, the goal of analysis, and the computational technologies we used.

1. **Chicken Diet**—This dataset was originally provided by Crowder and Hand [1990], in *Analysis of Repeated Measures* (example 5.3), published by Chapman & Hall. The dataset tracks the weight gain of chicks in four different diets: normal, 10% protein, 20% protein, and 40% protein replacement. There are 20 chicks who take the normal diet, and 10 chicks each for the remaining three so a total of 50 chicks. Weight for each chick was measured on the following days for three weeks: 0, 2, 4, 6, 8, 10, 12, 14, 16, 18, 20, and 21. Five chicks did not make it to the 21st day so the data contains missing values. The goal is to see how the weight of a chicken relates to a specific diet.

   Technologies used:

   - **Data preparation:** Ignore the days with missing values.
   - **Analysis:** linear and quadratic regression along with a root mean squared error metric.
   - **Statistics:** Unpaired significance testing, confidence intervals.

2. **Breast Cancer**—This dataset comes from digitized images of breast mass UCI. The features describe the cell nuclei in each image. The goal of the study was to classify tumors as benign or malignant using predictive modeling. Our reanalysis will test different machine learning methods to see which model gives the highest diagnostic accuracy and whether the difference in accuracy is statistically significant. In addition, the reanalysis will test the performance of each model with missing data to see which model can handle missing data the best. Finally, we determine which feature(s) are most important in making correct predictions.

Technologies used:

- **Data preparation:** reducing 30 cell-specific attributes into 7 useful features.
- **Data preparation:** impute missing data.
- **Analysis:** machine learning techniques including logistic regression, decision trees, random forests, and support vector machines.
- **Analysis:** the application of correlation to remove similar features in order to lower the dimensionality of the machine learning analysis.
- **Statistics:** precision, recall, and F-measure.

3. **Cystic Fibrosis**—The RNA-seq data for this analysis comes from NCBI GEO (GSE124548). The purpose of the original study was to study the effect of a drug (Lumacaftor/Ivacaftor) to treat cystic fibrosis [Kopp et al., 2020].

RNA-seq is a method for detecting the abundance of messenger RNA (mRNA) of each gene present in a sample drawn from a patient. Our goal in this book is to identify the genes whose expression might suggest cystic fibrosis. The main challenges have to do with normalization, multiple hypothesis testing, identification of significant changes, and use of confidence intervals to determine magnitude of change.

Technologies used:

- **Data preparation:** normalize RNA-seq data so that the RNA levels obtained from different samples are meaningfully comparable.
- **Analysis:** F-measure based on random forest prediction of diagnosis based on changed genes.
- **Analysis:** Random forest for the prediction of cystic fibrosis and importance ranking of genes.
- **Statistics:** Shuffle tests to determine the statistical significance of the differences in gene expression due to disease status and drug use. Multiple hypothesis testing correction.

CHAPTER 2

Chick Weight and Diet

2.1 GOAL OF THE STUDY

The goal of this study is to determine whether the diet provided to a young chicken can affect the chick's weight and, if so, by how much. We will compare the diets pairwise to see whether the differences in the final weights are statistically significant.

This form of analysis is common, viz. there is some controllable input (diet in this case) and some output of interest (final weight). We want to determine whether changing the input influences the output and by how much. A properly designed experiment will assign individuals (chicks) to one input group or another (one diet or another) based on random assignment (randomly and independently, always using the same probabilities, assign each chick to one of four diets). Because individuals (chicks) may have different final results for other reasons (e.g., some chicks may be genetically disposed to gain more weight than others), there will be variation within each group. Statistical analysis will tell us whether the differences between the diet groups are likely due to the diet or likely the result of random chance.

In addition to comparing final weights, we want to understand the relationship between a dependent variable (weight) and one or more independent variables (time and diet). We will use linear and nonlinear regression analysis for this purpose.

2.2 DATA DESCRIPTION

The chick weight dataset has 578 rows and 4 columns from an experiment on the effect of diet on early growth of chicks. The four columns are "Weight," "Time," "Chick," and "Diet." Weight is a continuous variable, time is ordinal (day since the experiment started), while chick and diet are categorical variables.

The dataset contains four diets: one normal diet and three with different amounts of protein replacements. The control group (normal diet, denoted Diet1) has 20 chicks, whereas the other diets each have 10.

You can find more information about the data here https://stat.ethz.ch/R-manual/R-patched/library/datasets/html/ChickWeight.html and it is described in Crowder and Hand [1990].

2.2.1   DATA PREPARATION

The body weights of the chicks were measured at birth and every second day thereafter until day 21.

Five chicks did not make it to day 21 so they were removed from the analysis. Here, we've chosen the most conservative approach to deal with missing data—remove the data items (chicks) having missing data. We can do this because there is ample data. Note that in an alternative study, we might be interested in the question of whether some diet increased the chances of mortality. But in that case, we would need far more chicks.

2.3   WEIGHT CHANGE

2.3.1   WEIGHT CHANGE SIGNIFICANCE

In order to determine whether changing the diet makes a significant difference to the final weight, we will use the shuffle test as described in *Statistics is Easy*. Since we have four different diets, we will compare the control diet, Diet1, against the three with increased protein content, Diet2, Diet3, and Diet4.

For our first comparison, Diet1 (the control) vs. Diet2, we can take the average of the final weights in each group and see whether the Diet2 weights are significantly greater than the Diet1 weights. This is a one-sided test, because we are interested only in weight increases in Diet2 relative to the control diet Diet1. In this case, the difference is 36.95 g. The question we are trying to address here is whether this difference might have arisen with a reasonably high likelihood (say greater than a 5% chance) just by chance.

To test this using a *nonparametric shuffle test*, we set a counter to 0 and then perform 10,000 computational shuffle experiments of the following form.

1. **Shuffle the diet labels among the chicks.** Here is what that means. Recall that a pair of vertical lines surrounding a group name, for example ||Diet1||, refers to the number of chicks in the group. Suppose there are ||Diet1|| chicks who received diet1 and ||Diet2|| who received diet2. Now consider the set of final weights of all chicks who received either diet1 or diet2. Shuffling has the same effect as choosing at random and without replacement the final weights of ||Diet1|| of the chicks in the two groups and putting them in group G1 and taking the rest and putting them in group G2.

2. **Evaluate the difference of the means of G1 and G2.** If it is as great as the difference observed (36.95 in this example), update the counter by 1. Otherwise, don't change the counter.

The p-value is the final (counter value/10,000). (If the counter value is 0, then we say the p-value < 1/10,000.)

Below is a histogram of the differences of all the shuffle experiments between Diet1 and Diet2 (Figure 2.1). The p-value is the proportion of shuffle tests that showed an average differ-

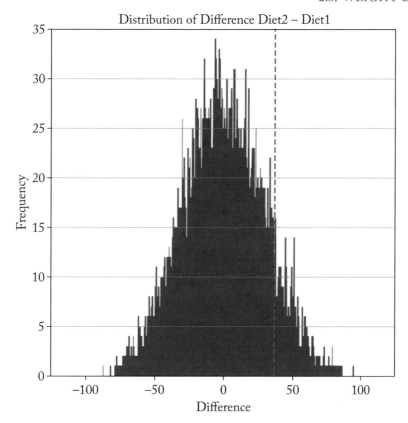

Figure 2.1: Distribution of the shuffled differences of the Diet1 and Diet2 chicks. The red line marks the observed difference between the diet groups. The p-value is the area to the right of and including the red line divided by the area of all the blue lines. 887 out of 10,000 experiments had a difference of means greater than or equal to 36.95. The p-value of getting a difference of two means greater than or equal to 36.95 is thus roughly 0.0887, which is not low enough to be statistically significant.

ence in weight equal to or greater than the difference observed in the study. Pictorially, that is the ratio of the area to the right of and including the red line (which is the observed difference) over the entire area of the blue lines.

Table 2.1 summarizes the results of the pairwise comparisons against Diet1.

The table shows that the Diet3 – Diet1 comparison has the smallest p-value, suggesting that the difference in weight is extremely unlikely to have happened by chance (roughly 0.07%). Diet2 – Diet1 also shows a difference, but has a roughly 8.9% chance of having happened by chance.

Table 2.1: Significance of different in weights in grams

Diets	Observed Difference	P-value
Diet2−Diet1	36.95	0.0887
Diet3−Diet1	92.55	0.0007
Diet4−Diet1	60.81	0.0074

Remark: As noted in the introduction, when multiple tests are performed, one or more tests could yield a low p-value just by chance (i.e., even if there were no effect). In Chapter 4, we discuss two corrections for this issue. The simplest and most conservative correction is called the Bonferroni correction. Effectively, it involves multiplying the p-values by the number of tests performed to get a "family-wise error rate." So, if we were interested in a family-wise error rate of 0.05, then Diet3 − Diet1 would pass (because 3×0.0007 is under 0.05) and similarly for Diet4 − Diet1.

2.3.2   WEIGHT CHANGE CONFIDENCE INTERVAL

For diet differences having low p-values, we next ask how big a difference we might expect to find. That is, if we were to collect data on different chicks for these same diets, what is the difference of the means that we are likely to see? To put this slightly more technically, for each diet D, we want to approximate a large number of random weight samples that are "similar" to the weights of the chicks who received diet D. Then we'll compare the samples from the different diets.

Mechanistically, we perform a new statistical experiment where we will sample, with replacement, from the original values of each diet separately to get a new set of values. This confidence interval calculation, called a *nonparametric bootstrap calculation*, goes like this.

To compare, say, diets1 and 3, perform 10,000 experiments of the following form.

1. Take the weights W1 of the ||Diet1|| chicks who received diet1 and form a group G1 of ||Diet1|| weights by taking ||Diet1|| values from W1 with uniform probability and with replacement. "Uniform probability" means that any of the ||Diet1|| weights has the same probability of being chosen. "With replacement" means that after a weight is chosen, it can be chosen again with its same probability. The net effect is that some weights in W1 may never appear in G1 and others might appear several times. The idea is to get a sample that approximates the final weights of an arbitrary group of chicks who get Diet1. Do the same thing to form G3 from the weights W3 of the ||Diet3|| chicks who received diet3.

2. Evaluate the difference of the means of G1 and G3.

Sort the 10,000 differences in ascending order (starting from the most negative if there are negative differences). The 90% confidence interval is the range from the 500th (the 5th

Table 2.2: 90% confidence interval of the differences in the mean weights for the different diets. For example, 90% of the time, we'd expect the mean of Diet3 chicks to be heavier than the mean of Diet1 chicks by between 50.29 g and 134.7 g.

Diets	Observed Difference	P-value	90% Confidence Interval
Diet3−Diet1	92.55	0.0007	50.29 .. 134.71
Diet4−Diet1	60.81	0.0074	29.05 .. 93.35

percentile) difference from the bottom of the sorted list to the 9500th difference (95th percentile) from the top of the sorted list. The table tells us that roughly 90% of the time, we'd expect the mean of a Diet3 group to improve on the mean of a Diet1 group by between about 50 g and 135 g.

The results are presented in Table 2.2. Note that we should compute confidence intervals only for those diets that have a sufficiently low p-value when compared with Diet1. As it happens, Diet3 and Diet4 have a statistically significant advantage over the control diet Diet1, but Diet2 does not.

Warning to the Unwary: A common mistake is to think that a 90% confidence interval that does not include 0 implies a low p-value (i.e., the result is unlikely to have occurred by chance). To see that this is not always true, consider a overly minimalist experiment in which one chick is given diet X and another chick is given diet Y. Suppose that the diet X chick weighs more at the end by 50 g. The confidence interval would be 50 g to 50 g (just the single number). But the p-value would be 0.5. This would thus be an utterly insignificant result that could have had nothing to do with the diet, e.g., the diet X chick might simply have had weightier genetics. One might conclude from this example that there simply needs to be some minimum number of data points (chicks in this case) in order to justify relying on confidence intervals alone. Unfortunately, this is not the case. While more data points tends to reduce the p-value when there is an effect, there is no fixed number to use. So, please take the time to do the significance test before measuring confidence intervals. If the p-value is high, then the confidence interval is meaningless.

2.4   REGRESSION ANALYSIS

Regression is a technique that is used to understand the numerical relationship between a dependent variable and one or more independent variables. For this analysis, we split the dataset into four sets where each set corresponds to one diet. For each diet, the independent variable is "Time" and the dependent variable is the weight of the chick. We will compare two regression methods (linear and quadratic) based on how well they fit the data (called "goodness-of-fit"). Note that another use of regression is to predict future data points (Figure 2.2).

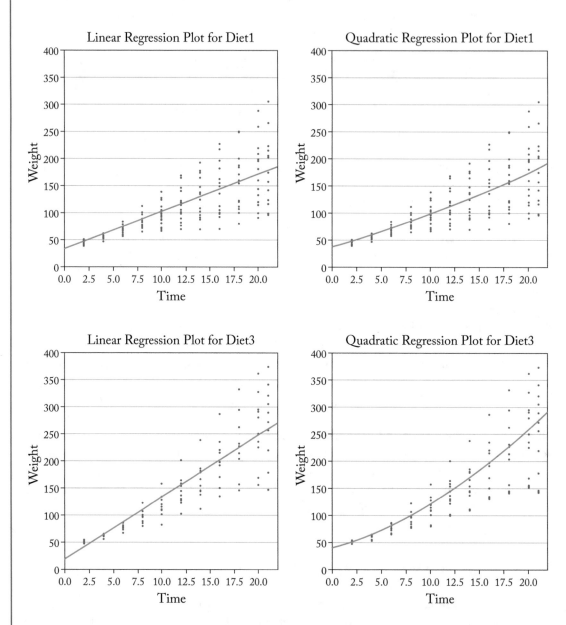

Figure 2.2: The linear and quadratic regression lines for diet1 and diet3. The quadratic regression lines curve upward toward the end, potentially giving a better fit.

Here we will use linear (degree 1) regression. We use the **polyfit()** function from the numpy package to calculate the slope and the intercept representing the rate at which the chick gains weight. These values are determined by minimizing the squared error.

```
# Polyfit gives the y-axis intercept,
# slope (for linear regression), and
# linear and quadratic coefficients (for quadratic regression).
# X and Y are the values for the two variables.
# Third argument is for the degree of the polynomial
# z now contains the coefficients for the polynomial.
import numpy as np
z = np.polyfit(X, Y, deg)
```

Using the polynomials and the original values used to fit the line, we can determine the squared error of the residual. A residual is the difference between the actual point (Y in the code) and the point on the fitted line (y_hat in the code). We calculate the squared error for each point. To calculate the RMSE (Root Mean Squared Error), we simply take the square root of the mean of the squared error (assuming n data points):

$$\text{RMSE} = \sqrt{\frac{\sum (Y - \hat{Y})^2}{n}}.$$

The RMSE for the four diets using quadratic and linear methods is summarized in Table 2.3.

To determine whether the difference between the linear and quadratic methods is statistically significant, we will perform a nonparametric paired swap test. A paired test is appropriate here since the same measurement points are being used for linear and quadratic regression.

Here is how to do this concretely. Build a table of four columns: chick, time point, error for linear regression, error for quadratic regression. Call that the original table, *orig*. Compute the RMSE of the third column and subtract from that the RMSE of the fourth column. This gives a value origDiff, which is the difference between the RMSE of linear regression and quadratic regression. Quadratic regression should give a lower RMSE (less of an error) because the regression line does not need to be straight.

Table 2.3: The diets have relatively small differences in RMSE (Root Mean Squared Error) between the quadratic and linear regression models. Diet3 shows the biggest difference, but even there the RMSE difference is under 5% of the linear RMSE.

Diets	Linear RMSE	Quadratic RMSE	Lin-Quad
Diet1	33.62	33.48	0.14
Diet2	41.01	40.69	0.32
Diet3	37.90	36.07	1.83
Diet4	20.39	20.17	0.22

```python
# This code evaluates the RMSE for both linear and quadratic regression.
# Using the z values, calculate the y_hat (the regression prediction)
# for each time value (x, below).
# Then calculate the Sum of squares ( Sum (y_hat - y)**2 )
def getSS(X,Y,z):
    all_diffs=np.array([])
    for i in range(len(X)):
        if len(z)==2:
        # linear regression z[0] is the slope
        # and z[1] the intercept
            y_hat = (X[i]*z[0]) + z[1]
        if len(z)==3:
        # quadratic regression
            y_hat = (X[i]*X[i]*z[0])+ (X[i]*z[1]) +z[2]
        all_diffs=np.append(all_diffs,[(y_hat - Y[i])**2])
    diff_sum = sum(all_diffs)


    return diff_sum

# Calculate the RMSE
def getRMSE(ss):
    diff_sum = sum(ss)/len(ss)
    return np.sqrt(diff_sum)
```

For every point in time for every chick, we will, with a probability of 50%, switch the linear regression and quadratic regression errors and recalculate the difference between the RMSE.

We will do this 10,000 times and count the number of times the difference was greater than the one observed. If this is rare, then the lower root mean squared error of quadratic regression is statistically significant.

```
# Pseudo-code for a paired test
# to compute the p-value of
# the difference in  RMSE between linear and quadratic regression.
counter:= 0
do 10,000 times
  create an empty table tmp
  for each row r of orig
    flip a fair coin
    if coin lands heads then
      insert r in tmp as is
    if coin lands tails then
      insert r in tmp with the linear regression error
      swapped with the quadratic regression error
  r3:=  RMSE of the third column of tmp
  r4:=  RMSE of the fourth column of tmp
  if (r3 - r4) >= origDiff
    then counter+= 1

p-value = counter / 10000
```

    To determine a confidence interval of that difference, we use the following bootstrap method:

```
# Pseudo-code for bootstrap to compute the 90
# of the difference in RMSE between quadratic and linear regression
# based on resampling with replacement the Diet1 weights.

do 10,000 times

    choose rows randomly and uniformly with replacement from orig
    to get a table of the same size as orig.

    Recompute the RMSEs of each column
    and recompute the difference.

Then sort the differences.
The lower end of the 90
500th difference and the upper end is the 9,500th difference.
```

Table 2.4: Quadratic regression does not have a statistically significantly lower Root Mean Squared Error (RMSE) than linear regression, but comes close in Diet3. The "NA" stands for "not applicable" which arises because the p-value > 0.05. When the p-value is high, the confidence interval has no meaning. Diet3 has a p-value close to 0.05, so the confidence interval might have meaning.

Diets	Linear	Quadratic	Lin-Quad	P-value	90% Confidence Interval
Diet1	33.62	33.48	0.14	0.27	−0.23 .. 0.51 (NA)
Diet2	41.01	40.69	0.32	0.26	−0.46 .. 1.14 (NA)
Diet3	37.90	36.07	1.83	0.052	0.00 .. 3.71
Diet4	20.39	20.17	0.22	0.26	−0.30 .. 0.72 (NA)

The p-value of obtaining a better fit using quadratic regression compared with linear regression is everywhere larger than 0.05, though it's close to 0.05 for Diet3 (Table 2.4). Recall that the confidence interval of a comparison is not meaningful and should not be computed when the comparison is not statistically significant. That is why we label the confidence intervals other than for Diet3 as "NA" meaning not applicable.

We conclude that there is no real benefit from a goodness-of-fit perspective to change the regression model from linear to quadratic. Different data (e.g., economic data showing the concept of diminishing returns) might gain substantially from quadratic regression.

2.5   EXERCISE

In the text, we studied the difference in Root Mean Squared Error (RMSE) between linear and quadratic regression and found that (i) quadratic gives a lower RMSE, (ii) the difference in their RMSEs is generally not statistically significant, except for Diet3 which has a p-value close to 0.05, and (iii) the difference is relatively small.

In this exercise, you will compute the 90% confidence interval of the RMSE with respect to the linear regression line L for Diet1. The goal is to gain an idea of how closely another set of chicks given Diet1 would track the original regression line L.

Hint: The new analysis will proceed as follows for Diet1.

```
# Pseudo-code for bootstrap to compute the 90
# of the RMSE with respect to  linear regression
# on the original Diet1 weigts
# based on resampling with replacement the Diet1 weights.

L := compute the linear regression line for Diet1
do 10,000 times
    construct a  bootstrap sample of chicks
    calculate the RMSE of  the time series of
    all the sampled chicks with respect to L
end do
compute the confidence interval  from the calculated RMSEs
```

2.6   CODE

A Python book with the code used for the analysis can be found here:
CaseStudies/Chick_weight_diet/Chick_Weight_Diet.ipynb.

CHAPTER 3

Breast Cancer Classification

3.1 GOAL OF THE STUDY

Here we will analyze the Breast Cancer Wisconsin (Diagnostic) Data Set.[1] The goals of this study are: (i) (diagnosis) to classify breast samples as cancerous or not; and (ii) (importance) to determine which measurement feature or combination of features is most important to determine whether a patient has breast cancer.

In many data analytic studies, one wants to determine which inference method performs the best. As an example, we compare four different classification methods: logistic regression, support vector machine (SVM), decision tree, and random forest.

The main technological challenges we will explore are:

- how to compare different machine learning models statistically (spoiler: random forests do very well when there are relatively few features, up to a few 10 s);

- the benefits of feature selection (spoiler: helps support vector machines a lot);

- how missing value decisions can affect accuracy (spoiler: imputation can work very well); and

- how to find the most important features (spoiler: machine learning methods will tell you).

3.1.1 APPROACH

For each classification method, we will split the data into a "training" set that we will use to build the model, and a "test" set that we will use to evaluate performance. We will then compare our predictions with the actual outcomes (benign or malignant). To evaluate robustness, we will repeat each analysis 100 times using different subsets of the data for training and testing, a process called "cross-validation." We will use these to compute confidence intervals for the F-scores in order to compare the quality of the results of the different methods.

The code for loading the data and doing the analysis is provided in more detail on the Github repository https://github.com/StatisticsIsEasy/CaseStudies.

---

[1](http://archive.ics.uci.edu/ml/datasets/Breast+Cancer+Wisconsin+(Diagnostic))

3.2   DATA

The features of the dataset are computed from digitized images of a fine needle aspirate (FNA) of breast masses. They describe characteristics of the cell nuclei present in the image.

3.2.1   ATTRIBUTE INFORMATION

1. ID number

2. Diagnosis (M = malignant, B = benign)

3-12. The mean of the following real-valued features are computed for each cell nucleus:

    (a) radius (mean of distances from center to points on the perimeter)

    (b) texture (standard deviation of gray-scale values)

    (c) perimeter

    (d) area

    (e) smoothness (local variation in radius lengths)

    (f) compactness ($perimeter^2/area - 1.0$)

    (g) concavity (severity of concave portions of the contour)

    (h) concave_points (number of concave portions of the contour)

    (i) symmetry

    (j) fractal dimension ("coastline approximation" $-$ 1), an indication of how ragged the nucleus is

13-22. Standard error of the 10 measurements listed above

23-32. Worst value for the 10 measurements listed above

The mean, standard error, and "worst" or largest (mean of the three largest values) of each feature were computed for each image, resulting in 30 features. For instance, consider the feature "Radius." The field 3 value is Mean Radius, the field 13 value is Radius standard error, and the field 23 value is Worst Radius.

In this study, for purposes of illustration, we will look only at the **mean** values.

3.3   DATA PREPARATION

3.3.1   TRAINING AND TESTING SPLIT

The data set consists of 569 breast samples, 357 benign and 212 malignant.

To construct our models, we will use 80% of the data for training and 20% for testing. We will call the features about cell nuclei used for prediction X, and the corresponding diagnoses

[B,M] (benign, malignant) Y. Mechanically, to split the dataset into training and testing, we will use the **train_test_split** function to split our dataset. The **test_size=0.2** parameter tells the method to keep 80% of the cases for training and 20% for testing. The **stratify** parameter tells the method to keep the same proportion of successes and failures (malignant/breast cancer and benign/healthy, for our example) in the training and test sets. We explicitly set this to **None** because the test set is supposed to represent new data and we want to put as few constraints on new data as possible.

```
# This code  performs the train-test
# (80
X_train, X_test, Y_train, Y_test = train_test_split(X, Y,
                                        test_size = 0.2,
                                        stratify=N)
```

As explained earlier, cross-validation is a method to get an average sense of accuracy across all the various data points for a given selection of hyperparameter settings. Here, we will randomly create the test and training dataset 100 times and build a distribution of the F-score using the default hyperparameter settings for each method. From the distribution we will take the 5th and 95th highest value which provide 90% confidence intervals for the prediction errors on training and test sets for each method.

```
# The code takes the features and labels as input and outputs
# the F-scores on the training and test sets
# using the logistic regression method.

score=np.zeros(shape=(100,2))
for i in range(100):
    X_train, X_test, Y_train, Y_test = train_test_split(X, Y,
                                        test_size = 0.2,
                                        stratify=N)

    score[i,]=(logreg(X_train, X_test, Y_train, Y_test))
```

3.4   CLASSIFICATION MODELS

A classifier is an algorithm that maps the input data (cell nuclei characteristics) to a specific category. We will test a few supervised binary classification models that assign every sample in the data set into one of the two categories of diagnosis, i.e., malignant (M, encoded as 1) or benign (B, encoded as 0). Thousands of machine learning models are possible, so if you have a favorite, you can substitute yours for ours in the code.

3.4.1   LOGISTIC REGRESSION

Logistic regression is a statistical model which, in its simplest form, is used to model how a binary feature (the outcome, here the cancer diagnosis) depends on features which may be discrete or continuous (for example the mean radius). Logistic regression differs from linear/quadratic regression, which forecasts a continuous dependent feature (e.g., a final weight in the chick example). Instead, logistic regression is used to infer a categorical/discrete dependent feature such as Benign or Malignant.

Logistic regression is implemented in Python in the scikit-learn library. Here's a small example:

```
# This applies logistic regression to make predictions
# and then calculate an
# F score (also known as F1 score) on both the training and test data.

from sklearn.linear_model import LogisticRegression
from sklearn.metrics import f1_score

def logreg(X_train, X_test, Y_train, Y_test):
    clf = LogisticRegression().fit(X_train, Y_train)
    Y_train_predicted = clf.predict(X_train)
    train_score = f1_score(Y_train, Y_train_predicted)
    Y_test_predicted = clf.predict(X_test)
    test_score = f1_score(Y_test, Y_test_predicted)
    return [train_score, test_score]
```

3.4.2   DECISION TREE CLASSIFIER

In a decision tree, each internal node represents the test on a certain feature, each edge is characterized by a result of that test. The leaf represents the category. In Figure 3.1, for example, the root node tests the value of the concave_points_mean feature. The leaves will give a diagnosis of Benign or Malignant.

A Decision Tree Classifier, as applied to this example, produces a decision tree that diagnoses patient conditions as Malignant or Benign based on the measured features. At each

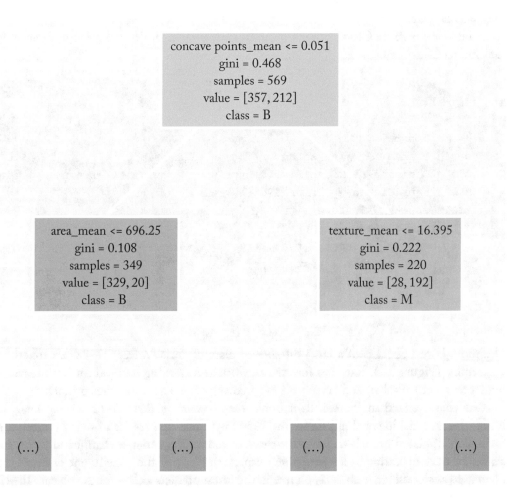

Figure 3.1: Decision tree classifier. The top line in each box represents the test on the data of a patient to perform. The second line *gini* is a measure of how effective the feature branching is in splitting the data. Of the 569 samples in the training set, branching on "concave_points_mean <= 0.051" assigns 349 patients to 0 (B for Benign) and 220 to 1 (M for Malignant). The blue node on the right is for the 220 samples that have a **concave points_mean > 0.051** and the orange node on the left is for the 349 samples that have **concave points_mean <= 0.051**. Out of the 220 in the blue node, a majority (192/220) of the samples are M (Malignant). The most informative further classifying feature-value for the 220 samples is "texture_mean <= 16.395" and the separation to the next level depends on whether the value is <= (left branch) or > 16.395 (right branch). The gray leaf nodes represent further levels of the decision tree.

node, the classifier tries to choose the attribute and the value that best separates Benign from Malignant cancers.

The Decision Tree Classifier is implemented in Python in the scikit-learn library under DecisionTreeClassifier class. Here's an example:

```
# This applies decision trees for prediction
# and then calculates an
# F score  on both the training and test data.
def decisiontree(X_train, X_test, Y_train, Y_test):
    clf = DecisionTreeClassifier(criterion="gini").fit(X_train, Y_train)
    Y_train_predicted = clf.predict(X_train)
    train_score = f1_score(Y_train, Y_train_predicted)
    Y_test_predicted = clf.predict(X_test)
    test_score = f1_score(Y_test, Y_test_predicted)
    return [train_score, test_score]
```

3.4.3 RANDOM FOREST CLASSIFIER

A Random Forest consists of a large number of distinct decision trees that work together as an ensemble (Figure 3.2). Random forests can work as a learning method for both regression (predict a numerical value) and classification (predict a class, e.g., cancerous or not).

As characterized in the excellent book *The Elements of Statistical Learning* by Hastie [2001], the essential idea behind Random Forests is to take the results of many noisy but approximately unbiased models and average them to reduce the variance. Empirically, this large number of trees operating as an ensemble outperforms any of the constituent decision trees. When used as a classifier, each decision tree in the forest predicts a class for the object. The class with the most (possibly weighted) votes becomes the prediction of the random forest model.

Here is an example of the implementation of Random Forest Classifier using scikit-learn library in Python.

```
# This applies random forests to calculate an
# F score  on both the training and test data.
def randomforest(X_train, X_test, Y_train, Y_test):
    clf = RandomForestClassifier().fit(X_train, Y_train)
    Y_train_predicted = clf.predict(X_train)
    train_score = f1_score(Y_train, Y_train_predicted)
    Y_test_predicted = clf.predict(X_test)
```

```
test_score = f1_score(Y_test, Y_test_predicted)
return [train_score, test_score]
```

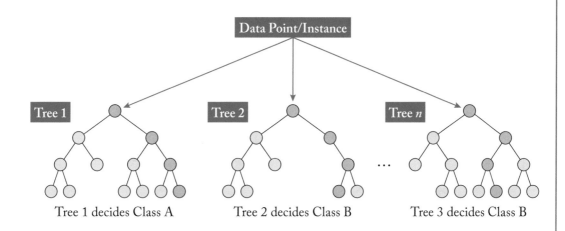

Figure 3.2: Simple random forest consisting of three decision trees. If the instance receives two votes for class B and one for class A, then the random forest will output B.

3.4.4 SUPPORT VECTOR MACHINE

Like random forests, Support Vector Machines provide methods for both classification and regression. The objective of the Support Vector Machine algorithm is to find a "hyperplane" (a linear equation among the features) that maximally separates the data points from different classes. The notion of maximal separation is expressed as a "margin" whose geometrical manifestation can be seen in Figure 3.3.

Here is an example of the implementation of Support Vector Machine classifier using the scikit-learn library in Python.

```
# This applies the support vector machine method to calculate an
# F score  on both the training and test data.
def mysvm(X_train, X_test, Y_train, Y_test):
    clf = svm.SVC().fit(X_train, Y_train)
    Y_train_predicted = clf.predict(X_train)
    train_score = f1_score(Y_train, Y_train_predicted)
    Y_test_predicted = clf.predict(X_test)
```

```
test_score = f1_score(Y_test, Y_test_predicted)
return [train_score, test_score]
```

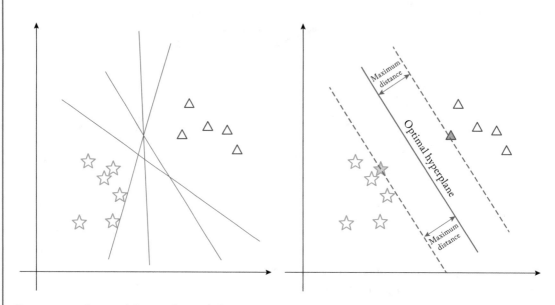

Figure 3.3: Optimal hyperplane. A hyperplane is a line when the data is two dimensional, a plane for three dimensional data, etc. Optimal in this example means choosing the line among all possible separating lines (examples of which are shown in the left panel) that separates the stars from the triangles with as large a distance (called the margin) on either side of the line as possible.

3.5   EVALUATION

Here we have combined precision (the number of those accurately predicted to have the disease/the number predicted to have it) and recall (the number of those accurately predicted to have the disease/the number who have it) through the F-score.

Recall that the formula for the F score is:

$$F\ score = 2\frac{(precision * recall)}{(precision + recall)}. \tag{3.1}$$

Table 3.1 shows the performance of the models that we've just discussed. (Because of the use of random seeds, your results may be slightly different.) We see that random forests perform best, which is in fact often the case when there are few features (on the order of a few 10 s). On

Table 3.1: The confidence intervals of the F-scores of different machine learning methods. The range reflects the 90% confidence interval of each method. Each method is run 100 times on a randomly chosen 80–20 split of the data. The 90% confidence interval is based on the fifth F-score in the sorted order of F-scores to the 95th F-score.

Method	Training F-score	Testing F-score
Logistic regression	0.86–0.89	0.82–0.94
Decision tree	1.00–1.00	0.84–0.94
Random forest	1.00–1.00	0.88–0.96
SVM	0.79–0.85	0.76–0.92

the other hand, when there are thousands of features, Support Vector Machines can perform better. We suggest trying the different methods on your problem as we are doing in this chapter. It doesn't take much work on your part and it can greatly enhance the quality of your results.

Note that, in Table 3.1, the methods show a larger confidence interval for the testing F-score than for the training F-score. The reason is that the test set is smaller, so there will be more variance in its F-score. This is illustrated in Figure 3.4. for the distribution of results from 100 logistic regression models.

```
# Plot the F-scores of logistic regression on
# training and testing data for different 80-20 splits.
sns.set(font_scale=1)
sns.distplot(score[:,0], label="train", bins=10, color="blue")
plt.axvline(score_sorted[5][0], 0,40, color="blue", ls='--')
plt.axvline(score_sorted[95][0], 0,40, color="blue", ls='--')
sns.distplot(score[:,1], label="test", color="green")
plt.axvline(score_sorted[5][1], 0,40, color="green", ls='--')
plt.axvline(score_sorted[95][1], 0,40, color="green", ls='--')
plt.xlim(0,1)
plt.legend()
plt.title("Logistic Regression")
```

Note to the Unwary: When Cross-Validation Is Insufficient

Because cross-validation splits the data into training (conventionally, 80% of the total data) and testing (20%) subsets, one might conclude that the result on the testing data is the result one

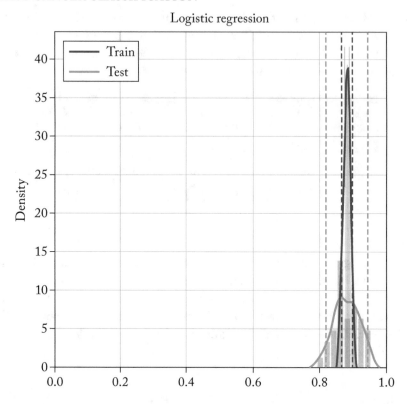

Figure 3.4: Distribution of training and testing F-scores for logistic regression. The vertical lines represent 90% confidence intervals. The test set size is smaller so will tend to have a wider confidence interval.

might expect if the model were applied to new data. However, this is true only if the analyst is very careful.

We, authors, have seen researchers deceive themselves into thinking they had amazingly great results, because they effectively incorporated test data into their training data. One way this happens is that one takes averages and/or other statistics on all the data and uses that to set some machine learning parameters. A second way is that one uses cross-validation with many different hyperparameter settings. A hyperparameter is commonly a configuration input to a machine learning model (e.g., number of trees used in a random forest) or a method used for handling missing data or some other method for handling data. If one runs cross-validation with different hyperparameters until one gets the best results, then one is effectively incorporating test data into the training process. We call that *polluting the test set*.

Polluting the test set will make the study results appear stronger, but then generalized application of the research in question will suffer worse results. That is just bad science.

To ensure that you avoid this, you would do well to sequester some of the input data from all other data at the very beginning. Then you can do cross-validation on the remaining data, optimizing hyperparameters to your heart's content. In the end, you create a model based on all the non-sequestered data and calculate an error on the sequestered data, which is the only error rate you should report.

In this case study, we use cross-validation alone for the sake of presentation simplicity (and do not optimize hyperparameters). That corresponds roughly to step (ii) of the following workflow: (i) sequester some data; (ii) perform hyperparameter tuning and cross-validation on the non-sequestered data to create an optimized model; (iii) apply the optimized model on the sequestered data; and (iv) report results only on that sequestered data.

Note to the Unwary: Imbalanced Datasets

How to measure error is also an issue. Our data set has the property that more records are marked B (benign) than M (malignant). This situation is called *class-imbalance* in machine learning parlance. Most real-world classification problems display some level of class imbalance. Usually, as in this case, there are more healthy/good s amples than sick/bad samples.

When there is gross imbalance (not our case), we may want to adjust our metrics. For example, if a disease occurs in only 0.1% of the population, then a classifier that always says the disease is absent will be correct 99.9% of the time. That would miss the entire point, however, because false negatives (i.e., claiming there is no disease when there is one) may be much more costly than false positives (i.e., claiming there is a disease when there is none). In such cases, we may want to put a larger cost on false negatives than on false positives. In the 99.9% example, we might want to assign a cost of 1 to a false positive (diagnosing illness when there is none) and 1000 to a false negative (diagnosing health when there is illness). For our breast cancer setting, we assign the same cost to false positives and false negatives, because there is relative minor imbalance.

3.6   MISSING DATA

Missing data occurs when some measurement wasn't done, was lost, or simply failed some quality check. We will show two basic methods for dealing with missing data: (i) removing records that contain missing values (as we did for chicks); and (ii) imputing (i.e., filling in) missing values.

Removing records having missing data is self-explanatory. We will simply remove all the records that are missing at least one value. Each record corresponds to a patient. This throws out a lot of information that may have been expensive and time-consuming to obtain, so we'd prefer an alternative.

The alternative is to impute the missing values with actual values based on some justifiable mechanism. One simple mechanism is to replace the missing values with the **median** value of the feature from the training set. For example, if the perimeter value is missing for a patient P, then assign the median perimeter value from all other patients to P.

```python
# This code  inserts missing data in row/column locations
# with probability portion_to_remove.
import random
def get_data_with_missing_values(data, portion_to_remove):
    data_copy = data.copy()
    ix = [(row, col) for row in range(data_copy.shape[0])
            for col in range(data_copy.shape[1])]
    for row, col in random.sample(ix,
                                    int(round(portion_to_remove*len(ix)))):
        data_copy.iat[row, col] = np.nan
    return data_copy
```

```python
# The first function removes rows (full patient records)
# that have any missing data.
# The impute function fills in missing data using
# the median method (taking the
# median value  for that measurement over all other patients).
def remove_missing_data_row(data):
    df = data.copy()
    df = df.dropna()
    return df

def impute_missing_data(data):
    data.fillna(data.median(), inplace=True)
    return data
```

To evaluate the robustness of various missing data methods, we also want to test the performance of the methods with different subsets of missing values. Below is a code example of using the *logistic regression* function with the *impute* strategy.

Table 3.2: 35% missing data: the second column shows the F-measures for each classification method on the test data when there is no missing data. The next two columns look at the 90% confidence interval of the F-measure on test data (i) when records having missing data are removed (ii) when missing values are imputed. Imputing gives better F-scores.

Method	Original	Remove 35%	Impute 35%
Logistic regression	0.82–0.94	0.73–0.90	0.79–0.92
Decision tree	0.84–0.94	0.67–0.92	0.71–0.86
Random forest	0.88–0.96	0.79–0.93	0.86–0.93
Support vector machine	0.76–0.92	0.00–0.84	0.73–0.78

Table 3.3: 50% missing data: even when 50% of the data is missing, imputing does remarkably well for Random Forests and Logistic Regression

Method	Original	Impute 50%
Logistic regression	0.78–0.93	0.79–0.91
Decision tree	0.84–0.93	0.74–0.90
Random forest	0.86–0.95	0.85–0.95
Support vector machine	0.79–0.91	0.74–0.88

```
# After imputing missing data using the median calculation,
# this code performs breast cancer classification
# using logistic regression.
score=np.zeros(shape=(100,2))
for i in range(100):
    X_train, X_test, Y_train, Y_test = train_test_split(X_select,
        Y, test_size = 0.2, stratify=None)
    X_missing = get_data_with_missing_values(X_train, 0.25)
    X_missing_imputed = impute_missing_data(X_missing)
    score[i,]=(logreg(X_missing_imputed, X_test, Y_train, Y_test))
```

The qualitative conclusion is clear: the accuracy of the machine learning methods is empirically less sensitive to missing data when using imputation than when removing records, especially for high levels of missing data (Tables 3.2 and 3.3).

3.7   FEATURE SELECTION

Feature Selection is the process of selecting or constructing a set of features in the hopes of obtaining better prediction results. In this section, we talk about the problem of selecting a subset of a given set of features and how that affects the prediction quality of each classifier (in the absence of missing data).

3.7.1   CORRELATION HEATMAP

A correlation matrix shows how the features are related to each other. If two or more features are highly correlated to each other, we may be able to keep just one and drop the rest as they are redundant. We use the **heatmap** function in the **seaborn** package to show the correlations among the features.

As you can see in Figure 3.5, several features are highly correlated with others. We remove the following features to get a reduced list of features on which to build our models (Figure 3.6):

- perimeter mean,

- area mean, and

- concavity mean.

3.8   IDENTIFYING IMPORTANT FEATURES

Scientists often want to explain some phenomenon by using a minimal set of features. This may reduce the number of tests that must be collected or simply give a more understandable picture of what is going on. While the goal of feature selection is to improve prediction performance by eliminating redundant features, the fundamental goal of identifying important features is to create a more understandable explanation, even possibly at the expense of prediction accuracy.

As an analogy, remember that first year physics started with the assumption of no air resistance in order to explain the best angle at which to throw a ball the greatest distance. The result might not have been as good as it would have been if you had taken air resistance into account, but ignoring air resistance gave an excellent first order approximation and a clear mental picture.

After removing the highly correlated features, all the classifiers were executed again. Table 3.4 summarizes the results.

All this said, it can sometimes even be a good strategy to use feature importance as a method of feature selection instead of using correlation. You will explore this possibility in the exercises.

Many classifiers, including the Random Forest classifier, give a ranking of the features that are most important to the determination of class membership. For this application, we see that the feature concave_points_mean dominates all others (Figure 3.7).

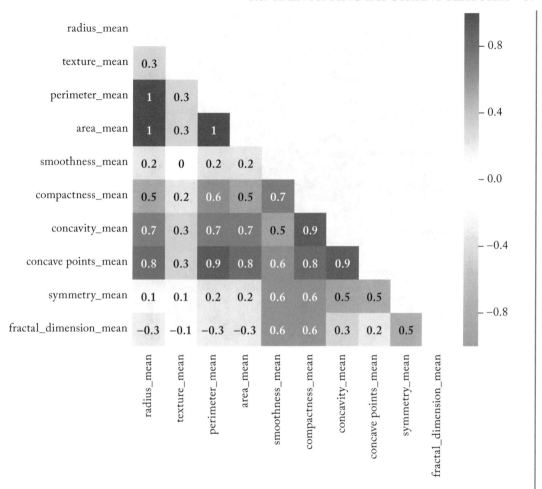

Figure 3.5: Pairwise correlation of the features in the breast cancer dataset.

Table 3.4: Feature selection improves the Support Vector Machine's (SVM's) testing F-measure but doesn't benefit the other models' F-measures. In our experience, tree-based learning methods don't benefit from feature selection unless there are at least hundreds of features.

Method	Training F-score	Testing F-score	New Training F-score	New Testing F-score
Logistic regression	0.86–0.89	0.82–0.94	0.84–0.87	0.80–0.92
Decision tree	1.00–1.00	0.84–0.94	1.00–1.00	0.83–0.93
Random forest	1.00–1.00	0.88–0.96	1.00–1.00	0.86–0.95
SVM	0.79–0.85	0.76–0.92	0.83–0.87	0.80–0.93

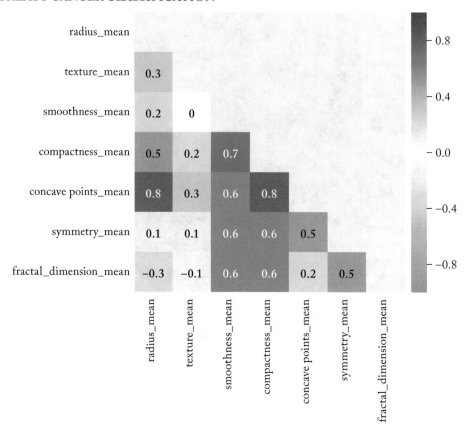

Figure 3.6: Pairwise correlations of the reduced set of features.

```
# This code uses a technique known as permutation importance
# to identify important features.
# We use 5 repeats but that is just the default.
from sklearn.inspection import permutation_importance
clf = RandomForestClassifier().fit(X_train, Y_train)
result = permutation_importance(clf, X_train, Y_train,
                                n_repeats=5, scoring='f1')
feature_imp_data = {'variables':X_train.columns,
               'mean':result.importances_mean,
               'sd':result.importances_std}
feature_imp_df = pd.DataFrame(data=feature_imp_data)
feature_plot = sns.barplot(x='variables',y='mean',data=feature_imp_df)
```

```
plt.errorbar(x=feature_imp_df['features'], y=feature_imp_df['mean'],
        yerr=feature_imp_df['sd']/np.sqrt(5), fmt='none')
feature_plot.set(xlabel="Features",ylabel="Mean Importance")
feature_plot.set_xticklabels(feature_imp_df['variables'],rotation=90)
```

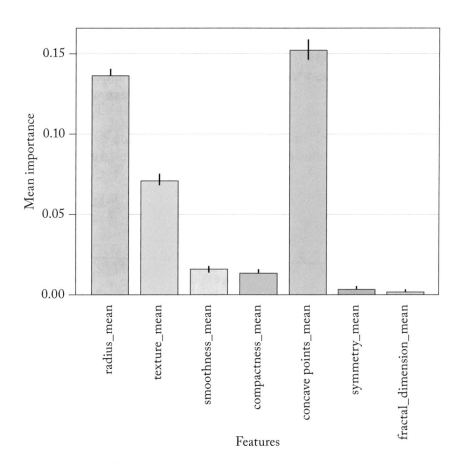

Figure 3.7: Feature importance on a scale of 0–1 on the vertical axis. The feature concave_points_mean is of primary importance in determining whether a patient has breast cancer. Recall that concave_points_mean was the most important feature in the Decision Tree which is why it was the feature evaluated at the root node of the tree. With more repeats, the error bars would be narrower.

3.9   EXERCISES TO TEST YOUR UNDERSTANDING

1. Imputing missing values as opposed to deleting them seems to improve the F-measure results for each of the machine learning methods. Evaluate the statistical significance of the improvement nonparametrically when 40% of the data is missing.

2. Imputation and random forests: Create a chart whose x-axis is the percentage of missing data (0%, 10%, 20%, 30%, 40%, and 50%) and whose y-axis contains the lower and upper bound of the 90% confidence interval for random forests using imputation.

3. Restrict yourself to the top k (e.g., $k = 5$) most important features and rerun the classification study using only those features. How is the F-measure impacted? If the F-measure does not decrease too much, the fewer feature model may be considered better based on the philosophical rule known as Occam's razor [Gauch, 2003].[2]

4. If someone presents to you the test error from cross-validation results along with the assertion that that error should be representative of errors on new data, what might you ask about the analytical process?

```
# Pseudo-code showing how to
# find the p-value in a paired # test setting.
Set counter = 0
Start with Tab
do 10,000 times
  tmptable = Tab
  for each patient,
    flip a fair coin
    if the result is heads
      swap the predictions between the M1 and M2 columns tmptable
   end for
  compute FM1-FM2 as X
  if X >= Diff then counter += 1
end do
The p-value is counter/10000
```

5. Redo the analysis done here (with whatever level of missing values you choose) by keeping out a subset of the data say 16% as sequestered. On the remaining 84% of the data, optimize

---

[2]The approach of retrying predictions using fewer features is sometimes called model simplification. If one simplifies by choosing various subsets of features to see which gives a good F-meausre, one is polluting the test set, because one is using the test set to determine the features to use. The net result is that one might get a poor fit on other data. Generalizability might suffer.

hyperparameters, missing value imputation, and anything else you choose on a series of cross-validation experiments. Then build a model with those optimized hyperparameter values on the 84% of the data and see how you do on the sequestered 16% of the data. Redo this process several times on a bootstrap of the sequestered 16% of the data (but with the same optimized hyperparameter values) to get a 90% confidence interval of the test set results. How does the test set result compare with the best cross-validation results on the 84% of the data?

6. Comparing the confidence intervals of the machine learning algorithms gave us some indication of which method was better, but did not establish statistical significance. Suppose that method M1 has an overall higher accuracy than method M2. To see whether that difference in accuracy is statistically significant, try a paired test. Compute the p-value to determine whether the difference in the F-measure of the random forest method compared to other methods is significant.

Hint: Here is how the paired test could work. (This is similar to what we did in the previous chapter to compare linear and quadratic regression.) Create a three-element table Tab: patient, M1 prediction, M2 prediction for some run of M1, and some run of M2. Suppose M1's F-score is FM1 and M2's F-score is FM2 and Diff = FM1 - FM2, where Diff > 0.

C H A P T E R   4

RNA-seq Data Set

4.1   GOAL OF THE STUDY

Cystic fibrosis (CF) is a genetic disorder that affects the lungs. In this case study, we will examine data from a study on the efficacy of a drug (Lumacaftor/Ivacaftor) to treat cystic fibrosis [Kopp et al., 2020]. This drug has been approved for individuals that are homozygous for mutations (i.e., the same mutation is inherited from both parents) in the CFTR (cystic fibrosis transmembrane conductance regulator) gene. Clinical studies have shown that the drug appears to be more effective with certain patients and less with others. This is likely because there are over 1000 known mutations in CFTR that are associated with CF.

The study researchers [Kopp et al., 2020] measured mRNA expression to help identify both the RNA-level differences between healthy patients and cystic fibrosis patients and between CF patients before and after drug treatment. Unfortunately, the published material doesn't reveal how the patients responded to the treatment. For that reason, the first part of this chapter simply asks whether we can detect any genes whose changes in expression are diagnostic of Cystic Fibrosis.

Brief Background on RNA and RNA-seq

Recall from elementary biology that a gene is a subsequence of the DNA of an organism. Each gene is activated or "expressed" when it creates RNA. Some RNA molecules are referred to as "messenger RNA" (mRNA, for short). The mRNA molecules of most genes are "translated" into proteins, which are the workhorses that carry out the instructions encoded in the DNA.

RNA-seq is a technology that counts the total number of RNA molecules in a sample. The data contains RNA-seq counts of mRNA molecules of each gene in the different samples. The samples will be groups of cells coming from either healthy patients, untreated cystic fibrosis patients, or treated cystic fibrosis patients.

Goals of this chapter:

We will use the data for two different goals.

1. Compare the healthy and cystic fibrosis patients to identify the genes whose expression differs significantly between the two groups.

2. Use machine learning to determine which of those significantly different genes are most indicative of sickness vs. health.

4.2   DATA

We obtained the RNA-seq data for this analysis from NCBI GEO (GSE124548). Blood samples came from:

- 20 healthy patients.

- 20 untreated patients with cystic fibrosis.

- The same 20 cystic fibrosis patients after treatment with Lumacaftor/Ivacaftor.

To analyze this dataset we will apply the following techniques.

1. **Normalization** — Researchers have provided the raw and normalized values. We will test if a simple normalization will provide the same results as the sophisticated one performed by the researchers.

2. **Unpaired tests** — When comparing two different populations (in this case, healthy vs. cystic fibrosis patients), we have to use unpaired tests. We did this also when comparing the final weights of the chicks, since each chick received only one diet.

3. **Multiple hypothesis testing/Reducing false positives** — When there are over 15,000 genes that could be expressed differently in healthy patients compared to sick ones, there might be differences in mRNA expression of some gene g just by chance. We would want to avoid calling such a gene g "differentially expressed," because that would constitute a false positive. So, we describe and use two multiple testing correction techniques: Bonferroni and Benjamini Hochberg.

4. **Evaluation of confidence interval** — Once we determine that a gene's expression is statistically significant different between healthy patients and cystic fibrosis patients, we will use confidence intervals to characterize the differences.

5. **Identifying diagnostic genes** — As input features for machine learning-based diagnosis, we seek the significantly differentially expressed genes (those that pass step 3) whose difference in expression between healthy and cystic fibrosis patients is greatest. Those will be genes for which the limits of the confidence interval of the difference in expression are both either highly positive or highly negative.

To avoid polluting the test set, we would have preferred to sequester some of the data from the post-normalization steps above to evaluate the models. We don't do that because we have so little data, so any conclusions we reach should be checked on disjoint data.

4.3   DATA PREPARATION

The researchers have provided the RNA-seq data in a database https://www.ncbi.nlm.nih.gov/sra?term=SRP175005. The first step in processing the raw sequence data is to determine the number of sequence reads that align to each gene, also referred to as the raw gene count. The authors have provided the raw gene count and the normalized values of the genes in NCBI's GEO database https://www.ncbi.nlm.nih.gov/geo/query/acc.cgi?acc=GSE124548.

We will begin the analysis by loading the Excel file from GEO's site where raw gene count and normalized gene expression data is provided.

```python
# Load the cystic fibrosis data.
import pandas as pd
data_df = pd.read_excel("https://ftp.ncbi.nlm.nih.gov/geo/series/\
GSE124nnn/GSE124548/suppl/\
GSE124548_AllData_170308_RNAseq_Kopp_Results.xlsx")
```

4.3.1   NORMALIZATION

The authors have provided raw and normalized data in their excel file, designated as RAW and NORM in the prefix of the headers. The normalized values are determined using the sophisticated methods of a widely accepted R package called DESeq2 [Love et al., 2014] that tries to estimate statistical parameters based on a distribution assumption. In the spirit of nonparametric methods, we will use a distribution-agnostic approach.

In the given dataframe, every column corresponds to a specific individual. Though the total number of mRNA sequences generated across individuals can vary, we are primarily interested in the relative expression of different kinds of mRNA. To make the data in different columns comparable, we will simply normalize the mRNA values from the genes of each individual (corresponding to one RNAseq run and one column) so their total read count is one million.

Simple Normalization Approach Code

```
# Cleaning and normalizing  RNA-seq data.

# Eliminate numbers in the name that are not relevant
# for ease of readability

data_df = data_df.rename(columns=lambda x: re.sub("w_[0-9]+_0", "w_0",x))
data_df = data_df.rename(columns=lambda x: re.sub("m_[0-9]+_0", "m_0",x))

# get all raw values
columns = list(data_df.columns)
raw_cols = [x for x in columns if "Raw" in x]

# calculate the total number of reads mapped to genes from each sample
raw_cols_sums = data_df[raw_cols].sum()

# divide each value by its column sum and multiply by 1,000,000
norm_cpm = [data_df[raw_cols].iloc[i]*1000000/raw_cols_sums \
            for i in range(data_df.shape[0])]

# keep a copy of data frame with gene name and description
norm_cpm_df = pd.DataFrame(data=norm_cpm)
norm_cpm_df = pd.concat([norm_cpm_df, data_df[columns[:9]]], axis=1)


# confirm results, columns should add up to 1,000,000
norm_cpm_df.iloc[:,:5].sum()
```

4.3.2   REMOVING GENES HAVING LOW EXPRESSION

Generally, genes that are expressed at very low levels (in the RNA-seq readout, that have low counts) do not have reliable measurement values. Furthermore, such genes may not be expressed at all in certain individuals. The authors of the original study have already eliminated some lowly expressed genes, but we will go one step further and remove any gene that has even one 0 value for some individual, because their expression levels are likely to be unreliable.

```
# Remove genes having any 0s in their counts.
number_of_zeroes = [ list(norm_cpm_df[raw_cols].values[i]).count(0) \
                   for i in range(norm_cpm_df.shape[0])]
genes_no_zeroes_logic = [number_of_zeroes[i] == 0 \
                   for i in range(len(number_of_zeroes))]
data_df_subset = norm_cpm_df[genes_no_zeroes_logic]
```

Removing all genes with at least one zero reduced the number of genes from 15,570 to 15,250.

4.4   DISTINGUISHING SICK FROM HEALTHY PATIENTS

When comparing samples from two different conditions, one should ask whether the measurements from the different conditions are paired. In our case we have three different conditions, (i) healthy, (ii) cystic fibrosis before treatment, and (iii) cystic fibrosis after treatment. The individuals constituting the healthy control group are disjoint from those who have cystic fibrosis. Therefore, their comparison will be unpaired.

In order to do the unpaired analysis, we will do a shuffle test for each gene. That is, for gene g, we start with by labeling each expression value as coming from a cystic fibrosis or healthy patient. Shuffling consists of shuffling the labels.

The net effect is the following: suppose that for some gene g, there are n1 expression values of sick patients and n2 values of healthy patients. Shuffling entails taking a random subset of size n1 from the n1 + n2 patients and labeling them "sick" and the remainder "healthy." If the health status of the patient does not affect the expression of gene g, we would expect the difference in the average expression of these randomly selected subsets to be about as large as we found when using the correct labels.

We are interested in identifying the number of times the shuffled comparison yields a more extreme value than the one observed. If often, then the p-value is high, suggesting that health status doesn't affect that gene's expression.

For each shuffle, we will calculate the *log-fold-change*. The *log-fold-change* is calculated by taking log base 2 of the ratio of the mean expression of those genes labeled S divided by the mean expression of those genes labeled H. The benefit of using log-fold-change is that it makes fold-increases and fold-decreases comparable in magnitude. For example, a doubling of magnitude corresponds to a log-fold-change of 1 and a halving corresponds to a log-fold-change of -1.

Here is a pseudo-code description of the shuffling process.

```
for each gene:
    observed change = calculate log fold change
    # log2( mean(cystic fibrosis)/mean(healthy))
    for each iteration: #do 100,000 times:
        shuffle healthy/sick labels
        associated with the gene expression values.
        randomchange = Calculate log fold change.
    Record number of times abs(observed change) <= abs(random change)
        into a variable "counter".
p-value for gene = counter / number of iterations
```

7,361 of the 15,250 genes showed significant p-values (< 0.05) Because we have performed p-value evaluations on multiple genes, some of those low p-values could have occurred just by chance. After all, a p-value of say 0.05 means that the observed change in expression had a 1/20th probability of happening by chance if the null hypothesis (in our case that cystic fibrosis had no effect on the expression of that gene) were true. Because we perform around 15,000 tests, this could give us around 300 (15,000/20) genes that look significant just by chance. Those would be false positives.

To correct for multiple hypothesis testing (in our case, testing many genes), we can use several different methods, but we consider just two widely used ones: (i) the extremely conservative Bonferroni correction [Bonferroni, 1936] and (ii) the looser Benjamin Hochberg's FDR (false discovery rate) correction [Benjamini and Hochberg, 1995].

4.4.1   BONFERRONI

The Bonferroni corrections limits the FWER (Family-Wise Error Rate). This error rate is the probability that at least one gene that is called differentially expressed is a false positive. In the Bonferroni method, we divide the threshold by the number of genes. So, if we take a family-wise error threshold of 20% or 0.2, the Bonferroni-corrected threshold would be 0.2/len(norm_p_values). In this case we have 0.2/15250 which is 1.3e-05. The only way a gene will pass this cutoff is if our shuffle test has 1 instance or fewer of obtaining a log fold change more extreme than the one observed in the 100,000 shuffles. There were 531 such genes. For those genes, the Bonferonni correction says that roughly 80% of the time, there won't be any false positives. Other researchers might choose different Bonferonni thresholds, but going much lower than 20% might result in no genes.

Table 4.1: Differentially expressed genes. More than 7,364 genes had p-values less than 0.05. Using a False Discovery Rate cutoff (Benjamani–Hochberg) of 5%, we get 6,082 genes. Using the FWER (Family-Wise Error Rate/Bonferroni) cutoff of 0.2 (so p-values of 0.20/15,250), we get 531 genes.

Total Genes	P-value < 0.05	FDR < 5%	Bonferroni 20%
15,250	7,364	6,082	531

4.4.2 BENJAMINI–HOCHBERG PROCEDURE

The False Discovery Rate (FDR), by contrast, is the percentage of genes that are predicted to be differentially expressed due to disease but whose differential expression is due solely to chance. Thus, FDR is a fraction of false positives whereas the Bonferonni correction asks whether there is at least one false positive.

The Benjamini–Hochberg method of creating a set of genes that have a given false discovery rate starts by listing the individual gene p-values in ascending order, from smallest to largest. The smallest p-value has a rank $i = 1$, the next smallest has $i = 2$, etc. Compare each individual p-value to its Benjamini–Hochberg critical value, $(i/m)Q$, where i is the rank, m is the total number of tests (15,250, as above), and Q is the false discovery rate the user can tolerate, say 5%. The gene associated with the largest p-value P that has $P < (i/m)Q$ should be included as should all of the genes associated with p-values less than P (including that aren't less than their Benjamini–Hochberg critical value). 6,082 of the genes having the lowest p-values passed this new cutoff.

Note that any gene that passes the Bonferroni criterion for a given threshold Q will pass the Benjamini–Hochberg criterion for Q. Here is why. Suppose that some gene g has a p-value P. By the Bonferroni criterion for some threshold Q, $P \leq (1/m)Q$, where m is the number of genes tested. For the Benjamini–Hochberg criterion, the threshold is that $P \leq (i/m)Q$. Since $i \geq 1$, the genes that pass the Bonferroni criterion must pass the Benjamini–Hochberg criterion (for the same Q) and will have the smallest p-values.

In our analysis we use $Q = 0.2$ for Bonferroni (meaning we are willing to consider a set of genes such that the probability that the expression difference of at least one of those genes could be due to chance is 0.2) and $Q = 0.05$ for Benjamini–Hochberg (meaning that we expect roughly 5% of the genes that pass the Benjamini–Hochberg test to be false positives) (Table 4.1).

In summary, the two methods say something quite different.

- A False Discovery Rate of x% says that roughly x% of the genes that Benjamini–Hochberg says are differentially expressed at the x% level will be false positives.

- A Bonferroni threshold of x% says that that there is roughly an x% chance that at least one gene that falls below the Bonferroni threshold will be a false positive. Put another way, there is a (100-x)% chance that there will be no false positives at all.

4.5   CONFIDENCE INTERVAL

Ideally, we would want to use just a few diagnostic genes to determine whether an individual has a condition. To help prioritize the significant genes, we look at the ratio of expressions, specifically the $\log_2$ of that ratio, and compute the 90% confidence interval of this log fold change. Our approach is similar to what we did for chick weights.

```
# Pseudo-code to compute the 90
# the log fold of gene expression.
for each gene that passed  the Bonferroni threshold:
    for 1000 iterations:
        bootstrap (sample with replacement) values
        from each condition and
        calculate the log fold change
    sort the bootstraps by log fold change
    report the 5
```

Given the confidence intervals of the differentially expressed genes, we rank them as follows. Suppose gene g1 has confidence interval $c1_{low}$ and $c1_{high}$ and g2 has confidence interval $c2_{low}$ and $c2_{high}$. Gene g1 has a higher rank (closer to 1) than g2 if $\max(|c1_{low}|, |c1_{high}|) > \max(|c2_{low}|, |c2_{high}|)$. Basically, this means that g1 has a more extreme log fold change than g2, either more strongly negative at the low end or more strongly positive at the high end of the confidence interval.

Of the top 10 most differentially expressed genes based on confidence intervals, the researchers [Kopp et al., 2020] discussed three genes in detail: MMP9, ANXA3, and SOCS3 (Table 4.2).[1]

4.6   PREDICTIVE MODELING

We will now evaluate the performance of the top 10 genes with the largest gene expression changes in predicting whether someone has cystic fibrosis. Cystic fibrosis can also be determined simply by checking for mutations in the CFTR gene. By contrast, our goal is to perform a diagnosis by measuring the mRNA from genes that are responding to the mutation. In the not-too-distant future, RNA-seq testing will become a routine procedure and therefore could test for multiple diseases without the need to sequence particular genes or to look for specific disease markers.

Similar to our analysis of the Breast Cancer dataset, we will determine which feature(s), in this case genes, are most useful in predicting the status of cystic fibrosis (Figure 4.1).

[1]In the R implementation of confidence intervals, MMP9 and ANXA3 are in the top 10, but SOCS3 is ranked number 16.

Table 4.2: Top 10 genes (lowest rank number), based on the 5% and 95% of the sorted log fold changes (the limits of the 90% confidence interval). The 10 genes after Bonferroni correction with the most extreme log fold changes happen to be over-expressed in cystic fibrosis patients, indicated by a positive log fold change.

Gene	Description	5% Value	95% Value
LOC105372		1.71	3.27
MCEMP1	Mast cell-expressed membrane protein 1	1.56	2.63
MMP9	Matrix metallopeptidase 9	1.61	2.50
SOCS3	Suppressor of cytokine signaling 3	1.38	2.22
ANXA3	Annexin A3	1.29	2.21
G0S2	G0/G1 switch 2	1.36	2.17
IL1R2	Interleukin 1 receptor type 2	0.92	2.11
PFKFB3	6-phosphofructo-2-kinase/ fructose-2,6-biphosphatase 3	1.40	2.11
OSM	Oncostatin M	1.27	2.03
SEMA6B	Semaphorin 6B	1.20	1.96

We have 40 samples, 20 healthy and 20 with cystic fibrosis. When we perform our analysis, cross-validation will use 20% as the test set, which is 8 samples, leaving us with 32 samples for training. It should be expected that there will be quite a bit of variation in the results depending on which samples are selected for training and testing. This was less of a problem for breast cancer because there were many more patients.

Warning to the Unwary: It's Easy to Pollute the Test Set

Before we start, we should elaborate on a problem with our analytical workflow that we alluded to earlier in this chapter. We have used the data D to determine which genes are differentially expressed in healthy vs. sick patients. Now we use that same data D to build a model to predict which patients are healthy and which are sick. This double use of the same data constitutes polluting the test set. What we would prefer to do is use one dataset D1 (of a subset of the healthy and cystic fibrosis patients) to determine differentially expressed genes and build a machine learning model on D1. Then we'd use a disjoint dataset D2 (of the remaining healthy and the remaining cystic fibrosis patients) to test the model. Unfortunately, we have too little data to do that here. So, this chapter can be viewed as performing the analysis on the D1 data. To conclude anything definitive, a careful analyst would test our resulting machine learning model on a disjoint dataset.

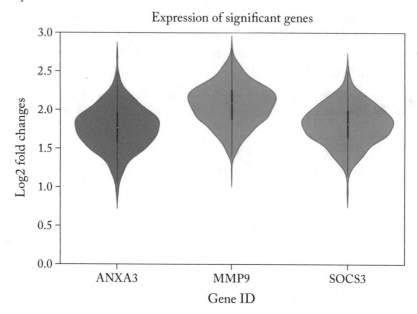

Figure 4.1: The figures show violin plots for the expression ranges of the three genes identified as potential markers of cystic fibrosis in Kopp et al. [2020]. These genes tend to have higher expression in cystic fibrosis patients than in healthy patients. Note that since this is based on log base 2 values, a value of 1 means that the expression is 2 times larger and a value of 2 means 4 times larger in a cystic fibrosis patient than in a healthy patient. A value of 0 would represent no change.

4.6.1 RANDOM FOREST INFERENCE

To determine whether we can distinguish healthy from cystic fibrosis patients based on gene expression, we developed a Random Forest model and tested it 100 times using an 80–20 training-test split. That is, each of 100 times, we take 80% of the data to train our model and test on the remaining 20%. The tests show that the model can diagnose patients with a precision, recall, and F-measure levels of nearly 90% (Figure 4.2).

The Random Forest method also allows us to determine which genes were most diagnostic. To find the most important genes, we construct a random forest using all 40 patients. Due to the random nature of the algorithm and the small number of data points, we will perform this method 100 times to get a distribution of the importance scores. The results are shown in Figure 4.3. The genes MMP9 and IL1R2 consistently show up as influential.

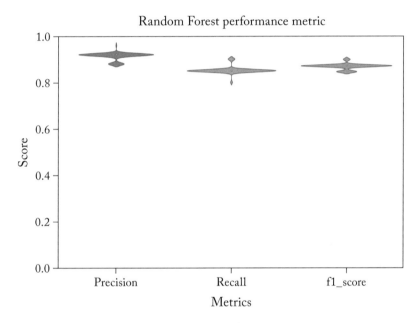

Figure 4.2: The figures show violin plots for the range of scores for three different metrics: precision, recall, and F-score. The mean value for the three metrics are 0.91, 0.85, and 0.87, respectively, suggesting that machine learning applied to differentially expressed genes may lead to accurate diagnosis, though one should apply the model to sequestered data to avoid polluting the test set.

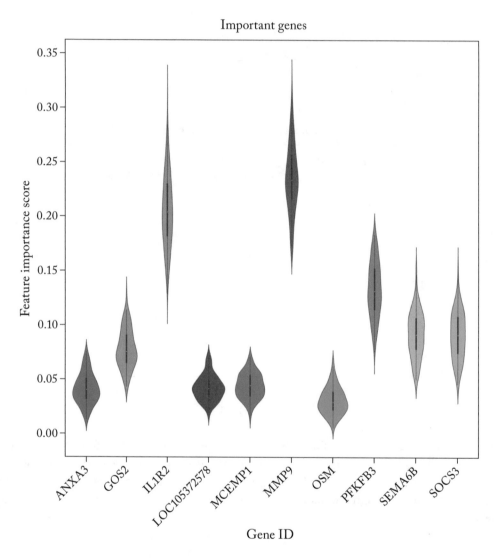

Figure 4.3: The horizontal axis shows the names of the most influential genes and the vertical axis shows their relative importance. Because the random forest makes strong use of randomness, the exact importance of any specific gene varies across different runs. Nevertheless, we see that MMP9 and IL1R2 are consistently important.

4.7   EXERCISES

1. Before studying which genes are most important in the random forest model, we should first show that the random forest predictions (diagnosing cystic fibrosis vs. healthy people) have a significantly better accuracy than randomly guessing the health of each person. Try that and determine the p-value.

   Hint: Remember that we have a dataset of 40 individuals, half of whom have cystic fibrosis and half do not.

   ```
   # Pseudo-code to evaluate the p-value of the null hypothesis
   # (i.e., random labeling is at least as accurate as
   # random forest labeling which has an F-score of 0.87).
    counter = 0
    Do 10,000 times
      Shuffle the labels on the patients
      Evaluate the precision, recall, and F-score
      If F-score >= 0.87
        then counter += 1
    End
    p-value = counter/10000
   ```

   The random assignment approach would assign a patient to cystic fibrosis or healthy with probability 0.5 which corresponds to the probability distribution of our sample. Then we would evaluate the F-score after that random assignment. The p-value is the number of times the F-scores resulting from the random assignment was greater than or equal to the F-score of the random forest (about 0.9).

2. We used the native random forest importance ranking to determine which genes were most influential in diagnosing cystic fibrosis patients. There is a widely used alternative method called permutation importance that we used in the last chapter. Describe what permutation importance does and then apply it to determine the most influential features.

3. We have used a random forest on the differentially expressed genes to try to predict which genes are the most influential in determining whether a patient is healthy or has cystic fibrosis. Then we evaluated the precision and recall of our analysis. Do the same analysis using support vector machines and compare the outcomes.

4. So far, we have done an unpaired comparison of untreated cystic fibrosis patients with healthy patients. We had to do this, because the two sets of patients were disjoint. By

contrast, the RNA samples after treatment were taken from the same individuals as before treatment, so a paired test (i.e., the gene expression value before treatment for a given individual X compared to the gene expression value after treatment for that same X) is appropriate. Paired tests often make it easier to identify changes to some given gene g due to treatment.

Perform a paired test comparing treated vs. untreated patients and see which genes show a significant log fold change and construct confidence intervals of that log fold change.

Rank the genes as we did before and see if you find any genes that were both differentially expressed in the healthy patient vs. sick patient comparison and in the sick untreated patient vs. the sick treated patient comparison.

5. Find a larger set of medical/genomics data. Sequester say s% of the data both from healthy and sick data. Then perform differential expression analysis and the diagnostic analysis (i.e., healthy vs. sick) on the remaining (1-s)% of the data, the "non-sequestered" part. On the non-sequestered data, optimize hyperparameter settings and anything else you choose on a series of cross-validation experiments. Then build a model with those optimized hyperparameter settings on the (1-s)% of the data and see how you do on the sequestered s% of the data. Without changing hyperparameter settings, repeat this process at least 100 times on a bootstrap of the initially sequestered s% of the data to get a 90% confidence interval of the test set results. How does the test set result compare with the best cross-validation results on the (1-s)% of the data?

CHAPTER 5

Summary and Perspectives

Processing scientific datasets involves three major steps: (i) data preparation, (ii) data analysis and inference, and (iii) statistical analysis.

The case studies in this book have included different methods of data preparation including normalization to render different value types comparable and imputation to estimate missing values. The case studies also presented a small collection of core machine learning algorithms from the library scikit-learn for inference purposes. Finally, the book presented several examples of the use of nonparametric statistics to determine the usefulness of an experimental condition (e.g., diet in Chapter 2), regression technique (also in Chapter 2), and machine learning method (in Chapters 3 and 4). We also considered methods to overcome the pitfalls of multiple hypothesis testing in Chapter 4.

Along the way, we discussed the possibility of polluting one's test set and how to combat this by sequestering test data.

Our hope is that you will use the code we provide to analyze your own datasets and perhaps even construct new Jupyter notebooks that we can link to from our github site.

We wish you the best.

Bibliography

Benjamini, Y. and Hochberg, Y. (1995). Controlling the false discovery rate: A practical and powerful approach to multiple hypothesis testing. *J. Roy. Statist. Soc. B*, 57:289–300. DOI: 10.1111/j.2517-6161.1995.tb02031.x. 9, 48

Bonferroni, C. E. (1936). Teoria statistica delle classi e calcolo delle probabilità. *Pubblicazioni del R. Istituto Superiore di Scienze Economiche e Commerciali di Firenze*, 8:3–62, Google Scholar. DOI: 10.4135/9781412961288.n455. 9, 48

Bushel, Pierre R., Ferguson, Stephen S., Ramaiahgari, Sreenivasa C., Paules, Richard S., and Auerbach, Scott S. (2020). Comparison of normalization methods for analysis of TempO-Seq targeted RNA sequencing data. *Front. Genet.*, 11:594. DOI: 10.3389/fgene.2020.00594. 4

Crowder, M. and Hand, D. (1990). *Analysis of Repeated Measures*, Chapman & Hall. DOI: 10.1201/9781315137421. 9, 11

Gauch Jr., H. G. (2003). *Scientific Method in Practice*, Cambridge University Press. DOI: 10.1017/CBO9780511815034. 40

Hastie, T., Hastie, T., Tibshirani, R., and Friedman, J. H. (2001). *The Elements of Statistical Learning: Data Mining, Inference, and Prediction*, New York, Springer. DOI: 10.1007/978-0-387-84858-7. 5, 28

Kopp, B. T., Fitch, J., Jaramillo, L., Shrestha, C. L., Robledo-Avila, F., Zhang, S., Palacios, S., Woodley, F., Hayes, D. Jr, Partida-Sanchez, S., Ramilo, O., White, P., and Mejias, A. (2020). Whole-blood transcriptomic responses to lumacaftor/ivacaftor therapy in cystic fibrosis. *J. Cyst Fibros.*, 19(2):245–254. DOI: 10.1016/j.jcf.2019.08.021. 10, 43, 50, 52

Love, M. I., Huber, W., and Anders, S. (2014). Moderated estimation of fold change and dispersion for RNA-seq data with DESeq2. *Genome Biology*, 15:550. DOI: 10.1186/s13059-014-0550-8. 45

Authors' Biographies

MANPREET SINGH KATARI

Manpreet Singh Katari is a Clinical Associate Professor and the Coordinator of Computational Studies in the Biology Department of New York University. In addition to teaching courses ranging from Statistics, Programming, Machine Learning, and Analysis of Next-Generation Sequencing Data, he also collaborates with researchers in the area of Plant Systems Biology.

His main passion is in developing software that empowers researchers to analyze, integrate, and visualize large-scale genomic datasets. Although his work has been primarily in the model plant species *Arabidopsis thaliana* he has applied his knowledge to many crops, such as Rice, Corn, Banana, and Cassava, and also to human disease datasets such as cancer.

SUDARSHINI TYAGI

Sudarshini Tyagi is currently a software engineer at Goldman Sachs where she uses machine learning particularly natural language processing and statistics to detect anomalies in financial regulations. She received her Master's degree in Computer Science from Courant Institute of Mathematical Sciences at New York University, where she wrote a thesis on visually detecting breast cancers from mammograms. She also holds a Bachelor's degree in Computer Science from Rashtreeya Vidyalaya College of Engineering, Bengaluru.

DENNIS SHASHA

Dennis Shasha is a Julius Silver Professor of Computer Science at the Courant Institute of New York University and an Associate Director of NYU Wireless. In addition to his long fascination with nonparametric statistics, he works on meta-algorithms for machine learning to achieve guaranteed correctness rates; with biologists on pattern discovery for network inference; with physicists and financial people on algorithms for time series; on database tuning; and tree and graph matching.

Because he likes to type, he has written six books of puzzles about a mathematical detective named Dr. Ecco, a biography about great computer scientists, and a book about the future of computing. He has also written technical books about database tuning, biological pattern recognition, time series, DNA computing, resampling statistics, and causal inference in molecular networks.

He has written the puzzle column for various publications including *Scientific American*, *Dr. Dobb's Journal*, and currently the *Communications of the ACM*. He is a fellow of the ACM and an INRIA International Chair.

Printed in the United States
by Baker & Taylor Publisher Services